The Wc
Lambing

Carol Puttock

First published in the United Kingdom in 2013
by The Sheep Lady

ISBN 978-0-9576239-0-3

Contents

Foreword

Whilst working as a shepherdess for many years, I have seen thousands of lambs being born, yet for me it is still an amazing thing to watch. Lambing is an exciting time, the beginning of a new cycle of life on the farm and a clear indicator of spring.

So many people ask me questions about what happens and what my role as a shepherdess is during this busy time, so I have written this book, using my own photographs and drawings, to answer some of those questions. I want to share with you this very special world, which I am so grateful to be part of.

Lambing indoors and outdoors are two very different systems, but both can be hugely productive. This book is based on my own experiences of indoor lambing systems on lowland farms, typical of those seen on many commercial sheep farms across the UK.

Some Sheep Facts

In Britain there are currently over 30 million sheep; that's about half the human population. Sheep have been bred for their meat, milk and wool for around 10,000 years. The domesticated sheep we have today descend from ancient breeds including the 'mouflon' sheep of Western Europe and Asia.

Sheep not only produce products that feed and clothe us, but they can significantly change how an area of grassland grows. Our beautiful island, with its soft hills and meadows, looks the way it does because of agriculture. And without sheep, the British countryside would be a very different place.

Sheep are light on their feet, will not damage or 'poach' the ground, can reach remote areas and fertilise soils with their dung, making them ideal for our varied landscapes. They are grazers and particularly good at eating back aggressive plant species, which allows a more diverse range of flora to grow. Having a greater variety of plants in an area attracts more wildlife and creates a more balanced habitat, beneficial to us all.

Sheep Characteristics

Sheep are prey animals with a flight instinct, grouping together and running when they are afraid. This sensitivity to fear gives us the impression that they are stupid, but it means life or death for them in the wild.

Sheep are generally perceived as unintelligent creatures, but I have found when given the opportunity, they can be quite the opposite, especially when kept in small groups with frequent human interaction. I have also found certain breeds to be brighter than others. Sheep need time to assess situations, they react well to calm energy and understanding how they think makes handling them much easier.

It is said, and I have observed this to be true, that sheep have very good memories and over the years build recognition of human and other sheep's faces. They develop strict pecking orders amongst themselves and on several occasions I have seen the same sheep leading the flock when on the move. They form close relationships with each other and get very stressed when they are alone.

To me sheep are inquisitive, fussy creatures that love routine and familiarity. You will notice throughout this book many of the ewes and lambs are looking straight at the camera. They are curious, even from an early age. I find it fascinating how they are so interested in us. This cheeky (some might say slightly scary) Shetland ram wanted to know what exactly I was up to.

This is me, looking after a friend's farm. I have done many things throughout my life. I've travelled the world and seen beautiful places, but my heart always brings me back here, to sheep. It is where I am content and can be myself.

When I tell people I am a shepherdess, they often ask me if I have my own sheep. Sheep are a huge responsibility and are very expensive and unpredictable to keep. So I work for other people on a temporary basis, helping mainly at lambing time on different farms. And yes, there are several lady shepherds about!

Shepherding as a job can take many forms. Some of my shepherding friends have their own flocks and work for themselves; they either own or rent grazing land for their sheep. Others are employed to work on a farm all year round with a salary. Some work on a contract basis, being paid per head of sheep they look after, whilst others, like me, are self-employed and help at busy times of the year.

What Hours Do We Work?

'A lot' is the answer to that! There is always something to do. Shifts tend to be 12 hours long, either in the day or throughout the night, and seven days a week. We often work many weeks without a day off.

Sometimes I tow my caravan to different farms to live in whilst I'm there. Generally lambing lasts for between three and six weeks on each farm. Some farms like their sheep to lamb during a short period of time, others prefer to take longer. The duration of lambing depends on when and how long the rams were put in to mate with the ewes.

Working on different farms and with different shepherds over the years has helped me accumulate the skills I have today. Getting to understand sheep, using and trusting my instincts and even observing all take practice. One of the best ways to learn about sheep is by working alongside good, experienced shepherds, and agricultural colleges will provide an all-round 'sheep education'.

Each year without fail I learn a valuable trick from someone. I'm always grateful for anything that might help me save a life. And the one definite thing I have become more certain of as the years have progressed is that absolutely nothing surprises me with sheep.

When is the Lambing Season?

Shepherds carefully plan the time of year a flock will give birth. There are many reasons for choosing when sheep will lamb and for each farm the needs will be different. Some lamb in the winter months but the majority, like this little Jacob, are born between March and May. The weather is warmer, the grass plentiful and full of goodness, ideal for ewes producing milk.

Trends in the agricultural industry fluctuate greatly and the sheep sector is no exception. There used to be many farms lambing from around Christmas time through to the spring, and shepherds like me had plenty of work during this time. Lambs were produced early in order to catch the Easter market and fetch a premium price. But over the last decade or so, running costs have increased substantially, and early lamb prices are not always worthwhile. Although some farms are still lambing early, many now tend to lamb later and outdoors, with much less financial outlay.

The last few years have also seen a general drop in sheep numbers, particularly after the foot and mouth crisis of 2001 – that is, until now. Wool is back in demand and the global requirement for meat has risen. Farmers are getting higher prices for their lambs and several I know have chosen to increase the size of their flocks.

Advantages of Indoor Lambing

One of the most significant reasons for lambing indoors is to shelter newborn lambs from exposure to bad weather. A high number of young lambs in this country die from hypothermia and lambing indoors can reduce this risk.

Lambs are also vulnerable to predators, particularly foxes and occasionally badgers. Keeping them inside for a couple of days until they are stronger can reduce these risks. And without sheep in the fields, the grass has time to rest and regrow.

Keeping ewes indoors means sheep can be penned up and kept in specific groups. In particular, ewes that have been ultrasound scanned can be split according to how many lambs they are carrying and then fed accordingly. Ewes expecting triplets, for example, can be fed more than those expecting just one lamb. Having scanned groups separated like this also helps us to foster more efficiently, which we will look at later.

Some shepherds keep all the first-time mothers together. These ewes can get pushed out at feeding time by the older ewes and generally take longer to have their lambs. They can be nervous and frequently require more attention.

Housing sheep indoors allows the shepherd greater control over the whole lambing process, and enables most problems to be dealt with quickly and easily. For me one of the greatest advantages of lambing indoors is that the sheep are so much easier to catch!

Disadvantages of Indoor Lambing

There are of course disadvantages to keeping sheep indoors for several weeks at a time. It is unnatural and although most breeds cope with being housed, others, particularly hill sheep, often do not and are better left outside.

The lack of fresh grass indoors can be a problem and result in nutrient deficiencies for some ewes. Sheep tend to get lazy when kept indoors for too long and often those lambing outside give birth with fewer complications, due to the constant exercise and lack of disturbance.

Another advantage of lambing outdoors is that when the sun goes to bed, so do the shepherds. Lambing indoors is much more intensive and keeping a shed staffed throughout the day and night is extremely expensive.

Keeping sheep indoors in such close proximity with reduced fresh air increases the risks of spreading disease. Eye infections, for example, are easily transmitted with sheep feeding so closely together. Keeping the pens clean with fresh straw is extremely important and something those lambing outdoors don't have to worry about.

A Farmer or a Shepherd?

This may sound a bit of an obvious question, but there is a difference between farmers and shepherds. Farmers have a wide range of knowledge in lots of areas such as arable and cereal crops or maybe livestock. They are generally good all-rounders. A shepherd, however, specialises in sheep, although most can turn their hand to a variety of jobs on the farm like fencing or harvesting when required.

These days it is said a farm needs to keep over 1,000 sheep to make employing one shepherd financially viable. That's quite a busy workload for a shepherd. The farms I work on keep between 500 and 2,000 breeding ewes. Depending on the size of the flock, the farmer or farm manager might employ a head and assistant shepherds.

Sheep Dogs

Unlike lambing outdoors, when a good set of dogs is often critical, I find lambing in a barn rarely requires a dog. Their enthusiasm to 'join in' with the fun is often very disruptive and jobs can take much longer. A ewe with new lambs will lose focus and panic with dogs around. Some breeds like the Mule and the Beulah are particularly protective and often chase after dogs to protect their lambs.

Dogs can, however, come in very useful, particularly when helping ewes who are not bonding with their lambs. The ewe's instinct is to protect her lamb and having a dog sit in front of the pen will strengthen her maternal tendencies. Dogs are also useful when it comes to holding back sheep whilst pen gates are opened, which may be done to put in hay or straw bales or perhaps at feeding time.

The dogs in the photograph (a mixture of collies and collie-kelpie crosses) were shut in the trailer, out of the way, whilst we turned ewes and lambs out into the field.

So Where Does the Lambing Cycle Begin?

It begins right here, with the ram. The ram or 'tup' is a critical part of the flock. He is responsible for the annual crop of lambs and much of the farm's income. This handsome ram is a Blackface or 'Blackie'. This popular breed is very hardy and thought to originate from Scotland.

It is often assumed that because a sheep has horns it is a ram, but this is not always the case. It can be the breed that determines whether a sheep has horns and not always the sex. The Blackface ewe, for example, also has horns.

Rams can be fertile from a young age, but are generally used for breeding from between one and two years of age, and continue up until about six or seven. One mature ram will be expected to mate with around 40 ewes during a typical month-long breeding season.

One question I am asked is whether it is true that rams are sometimes homosexual, and the answer is yes. Occasionally a ram can have no desire to mate with a ewe. Great, if you've just paid several hundred pounds for one!

Choosing a Ram

This photograph shows three common breeds of ram used in our commercial flocks: Charollais (left), Texel (right) and Suffolk (front). A ram is potentially a long-term investment and needs to be selected with care. Rams are often bred in specialised, pure-bred flocks, ensuring all best bits of a breed go into each animal. Buying in new animals with potential diseases is always a risk. Breeding your replacement rams at home will help reduce these risks.

As with all fathers both the ram's good and his bad genes will pass down to the lambs, so it is vital his body composition or as we say 'conformation' is healthy, hardy and fertile. The condition of his feet is also very important. If he is limping or has an ongoing ailment such as footrot, he will struggle to do a good job at mating or 'tupping'. It helps to prepare rams several weeks ahead of mating, to ensure they're in peak condition.

Prior to mating, rams are often fitted with a harness or 'raddle'. Attached to the raddle is a coloured crayon block which rubs off on the ewe's bottom when he mates with her. This raddle mark is an indication to the shepherd as to which ewes have been mated with and which haven't. Once the ram is ready to do his job, he will be put in with the ewes and nature left to take its course.

All Sheep Look the Same

Sheep can perhaps all look the same when you first start working with a flock, particularly if they are all one breed. But after a while, certain individuals become distinctive and more recognisable. The ewes in the photograph are known as North of England or North Country Mules. Looking at their faces, you can see their markings all differ slightly.

At lambing time we look at the same sheep many times during the day. Perhaps there is a distinctive marking on a leg or face. Sometimes the shape of the head or body is noticeable in some way. And sheep tend to lie repeatedly in a preferred place, making it easier to familiarise oneself with individuals.

Despite perhaps looking similar, sheep often have very different characteristics. One of the flocks I help to lamb in Sussex has 500 ewes. Some flock members are well-known characters. There is an adventurous one named Dora after Dora the Explorer, Biscuit, a sheep fond of digestives as a youngster, and one of my favourites, a mule called Jumper. She effortlessly hops from field to field as she pleases and carries that haughtiness sheep so frequently do.

Breeds of Sheep

There are many breeds of sheep to choose from. I personally believe choosing a breed you like, one that will fit into your farm, is really important. So too is selecting a breed to match the ground they'll be grazing on. Keeping hill or mountain sheep on rich lowland pastures doesn't always work, and similarly lowland breeds with healthy appetites would struggle on an exposed hill farm.

Sheep breeding is a very complex topic. Sheep breeds follow desirable trends and prices of breeding ewes can reflect this. Unlike rams, which are usually a pure breed, large commercial flocks are often made up of crossbred ewes. Crossbreeding means different qualities of certain breeds can be combined to create an animal which may suit a farm or situation better than a pure breed.

The North Country Mule, for example, is the creation of a Bluefaced Leicester ram mated with a ewe, usually a Swaledale. The Swaledale is a hardy hill sheep and great mother, but typically has just one lamb. The Bluefaced Leicester, however, is a lowland milky breed that can have lots of lambs. The product of these two breeds is a hardy animal with good mothering qualities which frequently has more than one lamb. In fact the North Country Mule is a very popular and successful crossbred sheep and you will see lots of them on our farms.

North Country Mule

Dorset

Suffolk x North Country Mule (Mule mother and Suffolk father)

Texel x Mule (Mule mother and Texel father)

Selecting Ewes for Mating

Sheep can live for many years, but commercial breeding ewes are usually kept until about six or seven years old. Their lifespan will depend mainly on three things: their body condition, their ability to produce milk, and their teeth, which if damaged or absent will prevent them from eating properly, consequently affecting their overall health.

Most sheep in the United Kingdom will lamb just once a year. Generally ewes are put to the ram at about 18 months of age, although some can be mated from eight months if they are physically big enough. When fully grown, the ewes I work with weigh somewhere between 70 and 90 kilos, but this depends primarily on their breed.

Once pregnant, ewes carry their lambs for five months or just under 150 days. Many farms ultrasound scan their ewes between 10 and 14 weeks after mating to determine how many lambs each individual animal is carrying. One ewe may be mated by several rams, and if the rams are of different breeds, the ewe can give birth to mixed breeds of lambs. Sheep have two productive teats on their udder and one ewe, in theory, has the ability to feed two lambs comfortably.

Many female lambs or 'ewe lambs' will be kept as future flock members. These sheep are often referred to as 'replacements'.

The Breeding Cycle of Female Sheep

Most breeds of sheep come into season during the autumn and winter months. The peak breeding season usually lasts from around October to December. Ewes are in oestrus (season) for about 30 hours every 17 days or so, and will only mate during this brief time.

As daylight hours shorten and nights become longer, an increase in the sheep's levels of melatonin (a chemical produced during darkness) sends a signal within the body which activates the reproductive system. Some breeds, however, can get pregnant outside the typical breeding season. The Dorset Horn is one such example, making it an ideal sheep for lambing out of season. Sheep breeds descending from or living near to the equator, where daylight hours are longer, have longer breeding seasons.

The sheep in the photograph were put with the ram in November and, as you can see, have blue or red raddle marks on their bottoms. Those with orange dots on their backs were scanned as carrying triplets.

If sheep were left to be wild most would still lamb in spring, but, like lots of things today, breeding cycles can be manipulated and controlled by us humans. Sheep breeding has really changed in the last few decades. For example, artificial insemination enables a ram's semen to be diluted and spread amongst many more ewes than he could serve naturally, or it can be stored for long periods. Occasionally embryo transfer is done: a ewe is encouraged to produce a lot of eggs at once, which are removed after fertilisation and placed into surrogate mothers. Sometimes rams will be vasectomised so that they can be placed with a group of ewes, encouraging them to come into season without actually making them pregnant.

What Do Sheep Eat?

The best and most natural food a sheep can eat is grass. Housed sheep have no access to fresh grass and can be fed supplements such as barley, oats or a concentrated pellet like those the sheep in the photograph are eating, in addition to silage or hay. Plenty of fresh water should always be available and a tub of minerals or a mineral lick will help them get all the nutrients they need to stay healthy.

Whilst ewes are carrying lambs, their energy requirements increase and even if they are still eating grass, supplementary feeding is usually necessary. Feeding correct rations is really important and overfeeding can be just as damaging as underfeeding. Food intake will influence many factors including ewe and lamb health, lamb size and milk production of the ewes.

Sheep are ruminants, meaning they have four separate stomach compartments. These compartments, with the help of micro-organisms, break down the grass and provide all the nutrition a sheep needs to be healthy. Sheep regurgitate and re-chew their food to aid digestion. This process is called 'chewing the cud'.

Waiting

Once all the sheep are indoors they can settle into barn life. It helps if they are brought in a couple of weeks prior to lambing, allowing them to adjust to the change of environment and lack of fresh grass in their diet. But for each farm circumstances will be different. Not all farms have enough indoor space and some will only bring sheep in just prior to giving birth. Some like to leave them out as long as possible to continue grazing, exercising and getting nourishment from the grass.

For shepherd and sheep it is a case of waiting. Like humans they can sometimes be a few days early or late in giving birth. Waiting can get quite boring if you are a shepherd and frustrating if you are the farmer paying the wages. But this is an ideal time to watch and spot any potential problems.

Ewes are kept together in big pens prior to lambing, like in the photograph. It is important to check they all come up to eat at feeding time. Those not interested in food could be ill and it is our job to monitor and treat any health problems.

Sheep lined up at the trough eating also provide a great viewing opportunity, with rows of bottoms on show. Any discharge or blood could be a sign of abortion or other potential problems, particularly in the week leading up to the start date. If I suspect there is a problem with a ewe, I will catch her and have a look.

Catching and Turning Over Sheep

When we need to catch a sheep, we can hook the leg or neck with a crook, or grab hold of the fleece. I avoid direct eye contact when I am catching one. If they recognise they are the target, they'll be off and catching them is very much harder.

Once they are caught, holding the head will stop a sheep from getting away. Turning the head and pulling against the back hip will gently take them off their feet and enable them to be put in a sitting or lying down position, as in the photograph.

It would be ideal if sheep could be left alone for the five-month duration of their pregnancy, but sometimes tasks need to be carried out on them. There are better times than others to work on pregnant ewes. It's ideal to leave them alone the few weeks just after tupping, when the embryo is attaching to the wall of the uterus.

At some stage during their pregnancy, ewes may be treated for the prevention of certain diseases. If they are limping, their feet will need attention, and many will be pregnancy scanned. Ewes might be sheared, or possibly dagged. (Dagging, also known as crutching, involves shearing away the wool around the bottom and tail to keep the area clean, as you will see in lots of photographs). Providing stress levels are kept to a minimum, these procedures can be carried out safely.

Just How Sensitive Are Sheep?

I've seen a sheep dream. A ewe was snuggled up with her lambs, fast asleep, chattering away to them. I watched her for a while and wondered if she was partly awake despite her eyes being closed. But then something made her jump and she woke up with a start. So I am convinced sheep dream.

I had a ewe once that craved motherhood. She had been tupped much later than the other sheep and wasn't due to give birth for several weeks. I noticed her lying, watching the fresh lambs intently through the hurdles. Eventually she stopped eating and became quite depressed, so I decided to move her right away from the lambing area. After a while she picked up and started to eat again. In a couple of months finally her turn came and she produced lambs. She was joyous to be a mother once again.

One year a human error occurred, which highlighted to me the sensitivity of sheep. A ewe was turned out into the field with twins, except unbeknown to us, one of the lambs didn't actually belong to her. The ewe's own lamb had got muddled up and was back in the barn with another mother. The mistake wasn't noticed until the next day. When the shepherd returned to the field to sort the problem out, the ewe was waiting by the gate. She knew she'd had two lambs and she knew one of them was back in the barn!

If sheep have a bad experience they remember it, and sometimes for years. For me, this is evidence of their sensitivity. Handling them with consideration, whether they are lambing or being sheared, wormed or loaded in a trailer, is, in my opinion, the least we can do.

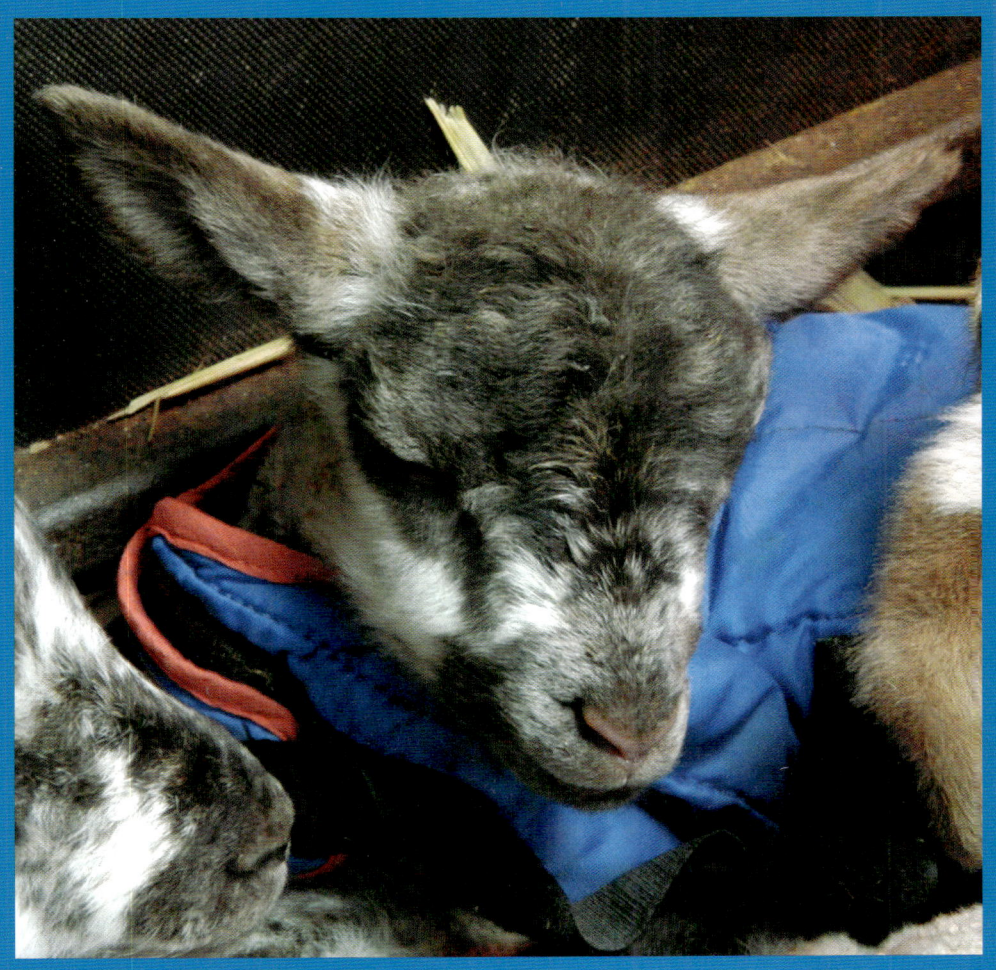

Ewe Illness at Lambing Time

Most shepherds will tell you a sheep's favourite occupation is getting a disease and then probably dying from it. There are many illnesses and ailments sheep get; some are easy to diagnose, others are not. Ensuring sheep are healthy and have a sufficient diet goes a long way in reducing their susceptibility to disease. Certain diseases occur through environmental circumstances or at specific times of year. We learn what signs to look for and how to recognise and treat sick sheep.

Many people ask if we use vets much. These days we tend to work with vets rather than calling them out on a regular basis. Usually amongst the staff on the farms we have enough skill and experience to overcome most lambing problems. If we are unable to get a lamb delivered for whatever reason the vet will perform a caesarean, but thankfully these situations are uncommon. If we repeatedly get a problem we're unsure about, we will consult our vet and work on a solution together.

Death is very much a part of farm life and although we try our best to keep everything alive and healthy, there are inevitably losses. Particularly at lambing time we have to anticipate losing a few ewes and no matter how hard we try, lambs will die. Whilst the animal's welfare is always paramount, we have to remain realistic about the economic consequences of prolonged or expensive treatment. Sometimes, if an animal is unlikely to recover, we will take the decision for it to be humanely destroyed.

There really are so many conditions sheep are prone to and it seems overwhelming when starting out as a shepherd. By encountering these problems, we learn to recognise and deal with them. Next I will explain some of the conditions we are likely to see around lambing time.

Diseases causing sheep to abort. Abortion in pregnant sheep is quite common and a potentially serious problem within a flock. There are several diseases which cause sheep to abort, many of which can be passed on to humans, which is why pregnant women are advised to stay away from sheep. (Transmissible diseases which pass from animals to humans are known as zoonoses.) Causes of abortion include Salmonella, Toxoplasma, Chlamydophila and Campylobacter organisms. These may be spread by direct contact between sheep or spread between flocks by other animals, even birds. Depending on the type of disease, lambs may be aborted or stillborn. Some become 'mummified' and are delivered together with normal healthy lambs.

Certain preventative measures including vaccination can help reduce the occurrence of abortive diseases. Breeding your own sheep and keeping the flock 'closed' as we call it, where sheep from other farms are never introduced, can prevent the introduction of many diseases including those that induce abortion.

Parasites. It is normal for sheep to have a small number of intestinal parasites, or gutworms. An adult ewe's natural immunity keeps these under control except around lambing time when her immunity system becomes temporarily depressed. To prevent this from causing problems, many ewes are wormed at lambing time, which helps to stop the spread of the gutworm eggs onto the pastures.

Cervical prolapses can frequently affect ewes at lambing time. As you can see it looks like the ewe has a pink bubble coming out of her bottom. Prolapses can happen many days before a ewe gives birth. The pressure inside her tummy builds as the lambs grow in late pregnancy and the muscles inside her become weak. There are several factors thought to contribute to this condition, including age, diet and the vulnerability of certain breeds. As you can see, it doesn't put her off eating. The shepherd will put a harness or similar device on her to keep the prolapse in place, allowing her to urinate normally and preventing the condition from deteriorating.

Prolapse of the uterus is something else we might see, but thankfully not very often. A prolapsed uterus occurs after the ewe has given birth. The whole uterus (still attached to the ewe) is pushed out after the lambs have been born. It is said an infection or excessive straining to get rid of the placenta (afterbirth) can cause this condition. It is quite incredible how large the uterus is. When it comes out, it looks as if it will never go back in, but with very much care it can be put back in place, allowing the ewe to make a full recovery and rear her lambs. This can be left to the vet, unless the shepherd is very experienced. These ewes will not be kept for breeding in the future.

Calcium deficiency or hypocalcaemia is a shortage of calcium in the ewe's bloodstream. It can occur in late pregnancy or shortly after giving birth. It is similar to milk fever in cows. A hypocalcaemic ewe will become weak and wobbly. Once we've identified the problem, we'll inject her with a calcium solution. Ewes mostly respond very quickly to this treatment and it is possible to save them if the condition is caught and treated early.

Twin lamb disease or pregnancy toxaemia can occur in late pregnancy, if a ewe's dietary intake becomes inadequate and her body challenged by the increasing requirements of the growing, unborn lambs. Ewes may lose their appetite, go blind and separate themselves from the others. Symptoms are often similar to hypocalcaemia. We orally administer a high-energy, rehydrating liquid at intervals. These ewes often take longer to respond to treatment than calcium-deficient ewes and much care is needed with them over several days if they are to survive.

Magnesium deficiency or hypomagnesaemia, sometimes called 'grass staggers', affects ewes mainly when they are turned out onto grass after lambing. Magnesium is not readily stored in the body and ewes need to take it in on a daily basis. I was taught by a shepherd that cold wet spells or frosts reduce the ewes' ability to absorb magnesium and the prevalence of magnesium deficiency within the flock increases. Sadly these ewes die very suddenly and often we cannot save them in time. There are preventative measures that may be taken if farms have a specific reoccurring magnesium deficiency problem.

Mastitis in sheep is an unfortunately common bacterial infection of the udder. It usually occurs when ewes start producing milk in the first few weeks after giving birth. The udder swells and becomes very hot from the infection. The milk can be clotted and show signs of blood. A major cause of mastitis is the transmission of bacteria from a lamb's mouth to the teat. Both dirty bedding and dirty hands will also help spread the disease and we often see more cases after a cold spell.

In order to protect what must be a very painful udder, ewes walk stiffly on their back legs and persistently move away to prevent lambs from sucking the teat. In some cases they become very sick, very quickly. They have a high temperature and the prognosis is usually not good.

If caught early, the condition can be treated by the shepherd with a course of antibiotics and anti-inflammatories. This will usually restore the ewe's health, but she will most likely never produce milk from the affected half of her udder again. More severe cases can result in the loss of the udder through gangrene, and sometimes ewes will die.

Retained afterbirth. Each lamb is attached by an umbilical cord to its own placenta or afterbirth, which is how it feeds and develops throughout its time in the uterus. Once the lamb is born the afterbirth will come away naturally from the ewe in a few hours. If for any reason it doesn't pass out freely, we give her antibiotics to reduce any infection. Ewes often eat their afterbirth: yes, revolting, I know. Some say ewes do this to hide any evidence to potential predators that a lamb has been born, whilst others say it is for the nutritional value.

Winter Shearing

The shorn sheep in this photograph and throughout this book are from the Overbury Estate in Gloucestershire. The Estate tried pre-lambing shearing in 2012 for the first time. Most breeds do not shed their wool naturally and will need shearing once a year, usually in the early summer. Winter shearing is not particularly common in the UK, but it is becoming more popular as there are several benefits.

Shearing ewes with lambs at foot can be stressful for all, and shearing is much easier done before lambs come along. Turning out sheep with short fleeces in the spring is also highly beneficial in avoiding flystrike, a very nasty and potentially fatal condition that affects sheep with fleeces. Shearing also prevents heavily fleeced animals getting stuck on their backs and dying.

Without fleeces, a lot more sheep can be fitted into a shed, and it is said unborn lambs will be more likely to go full-term and be born stronger to shorn sheep. Shepherds can clearly see the condition of sheared ewes and any pre-lambing problems. Lambs can find the teats easier and the risk of lambs being suffocated under the mother's heavy fleece is avoided. The shorn sheep at Overbury were also seen to take their lambs to shelter during bad weather, whilst the unshorn sheep did not.

Although combs used to shear sheep in the winter will leave on more fleece than summer combs, cold weather and shorn sheep is not always a good combination and in my opinion winter shearing is not suitable for every flock. But providing ewes are housed and turned out when they have a reasonable fleece growth, like the sheep in the photograph, winter shearing can work well.

Is She Pregnant?

There will almost certainly be a few ewes in every flock that will not be pregnant when it gets to lambing time. In healthy flocks, I would say less than 2 in 100 will, as we say, be 'empty'. Occasionally ewes can simply be incapable of pregnancy or 'barren'. Others will have conceived, but varying factors may have caused one or more foeti to 'slip' or be 'reabsorbed'. The ewe's diet may have been compromised at some stage of the pregnancy, or she may have contracted a disease causing her to lose lambs. Young ewes may be given another chance to breed the following year, but if a sheep continually fails to produce a lamb, it will be sent to market as a 'cull' (a ewe that has reached the end of its productive life).

By looking at a ewe's udder or 'bag' we can tell if she is pregnant. Comparing the two ewes in the photograph, we can see the one on the right has a developed udder-holding milk. As we say, she is 'bagging up'. She also has a swollen pink bottom, telling us she is not far away from giving birth. The ewe on the left, however, has no udder development and is very unlikely to be pregnant. The lack of a raddle mark suggests she did not come into season or mate with the ram.

How Do We Know a Ewe Is Starting to Give Birth?

Once the lamb is ready to be born, several hormonal changes begin to happen to the ewe and continue over the course of many hours. During the final stages of the process, she becomes very restless. She will choose a particular place in the barn, often away from the others, mostly in a corner or against a wall like the ewe in the photograph.

She will stand up and lie down repeatedly, going round in circles and pawing at the ground to dig herself a nest. This can go on for a good hour or more.

She might keep bleating and will usually hold her tail in the air. Sheep generally stop eating and chewing the cud when they are getting ready to lamb and, as we can see in the photograph, look up or 'stargaze' as the contractions start.

Which Way Should a Lamb Come Out?

This is ideally how a lamb should be born, with head and front legs together. In real life, there is very little room to manoeuvre inside a ewe.

This is a breach birth, when the bottom comes out first and the legs are tucked back inside the ewe. These lambs can be tricky to deliver.

Then there are backwards ones. These lambs often need delivering as they can get stuck half way out and risk drowning if the umbilical cord breaks too soon.

Lambs born head first need delivering. The head comes out, whilst the body stays inside. The head will swell up and the lambs will die if they are not delivered.

Ewes having several lambs occasionally need help if lambs get muddled up inside.

And lambs coming out like this, with one leg and a head first, might need help too, although ewes can sometimes manage to push these out by themselves.

The Water Bag

O nce the contractions have begun and the ewe starts pushing, a bubble with the amniotic fluid appears out of her bottom. This is the water bag which has surrounded and protected the unborn lamb inside her. When we see the water bag, we know the birthing process is underway.

A very common question I am asked is 'how long does it take for a sheep to give birth?' All sheep are different and each one will give birth at a different speed. The size of the lamb and how it is presented in the birth canal will determine how quickly it is born. Single lambs, for example, can be big and will often take longer than twins or triplets to come out. Ewes lambing for the first time are usually slower to give birth, whereas older experienced mothers can literally pop one out in a few minutes. Generally, we would expect to see a lamb emerging from a ewe within an hour after the water bag has come out.

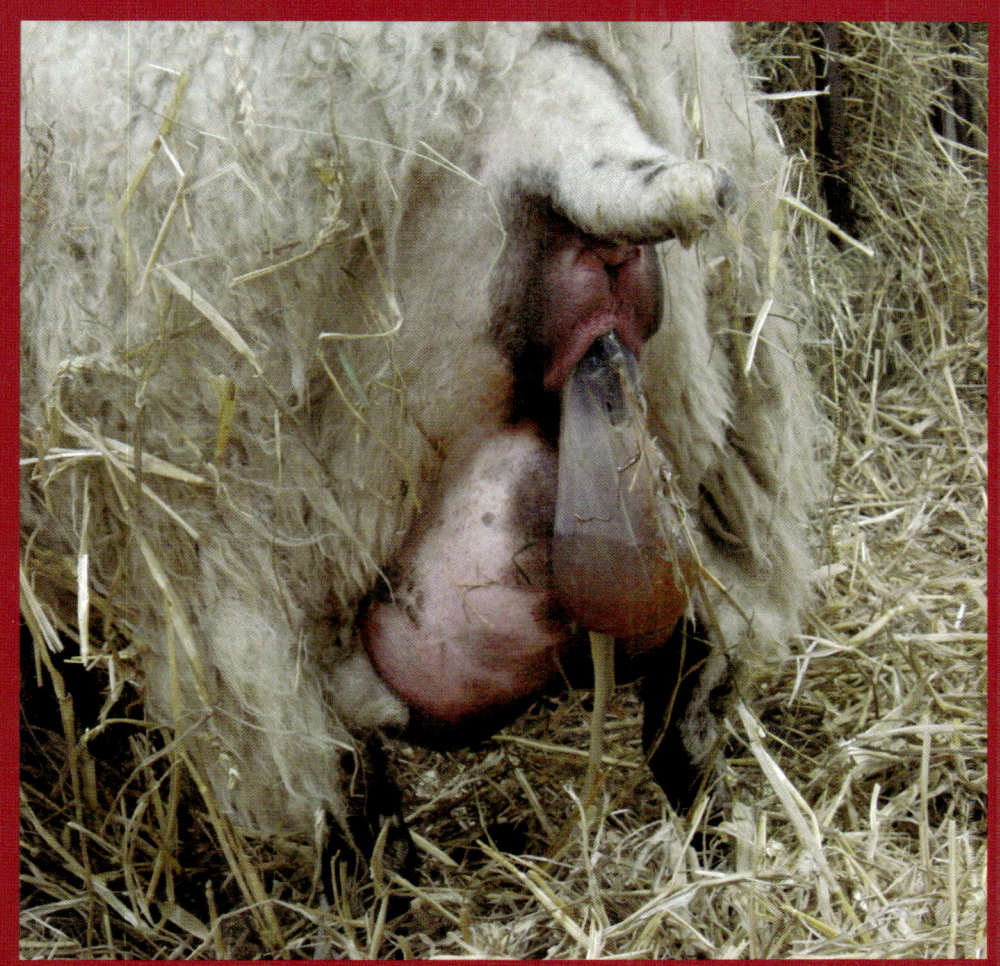

A Natural Birth

After the water bag has come out, the lamb's front feet and nose begin to appear:

... and then a bit more.

After 25 minutes a healthy lamb is born.

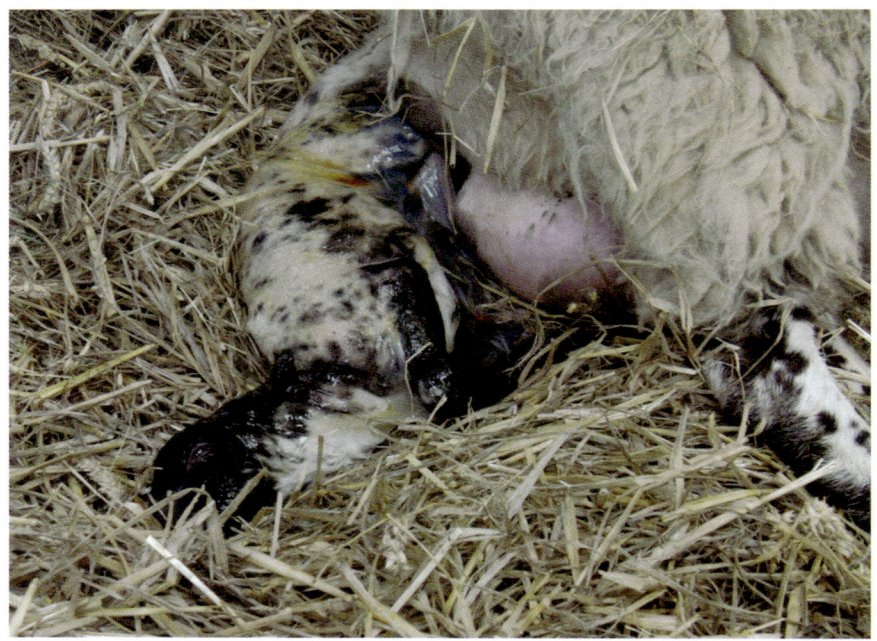

As the ewe stands up, the umbilical cord breaks away.

How Long Will It Take a Lamb to Stand Up?

If lambs have a natural birth without any trauma and have been born to a healthy ewe, they will generally stand within about ten minutes. This little fellow is beginning to get to his feet. As you can see the ewe is licking him dry; this stimulation also motivates him to get up.

In the wild it is vital freshly born babies stand up as quickly as possible to avoid being eaten by predators. This instinct still exists in the domesticated sheep world. Some breeds will get to their feet much quicker than others and some farms will choose certain breeds for this very quality. The Charollais cross lambs I work with, like the one in the photograph, are particularly good at getting up quickly.

A lamb's health will reflect the health of the ewe. Unhealthy sheep which have been fed inadequate diets or may have been unwell will more than likely produce weak lambs with possible nutrient deficiencies. These lambs can take much longer to get going.

How Do We Know There Is a Problem with a Birth?

I am frequently asked how we know when a ewe is having problems giving birth and at what stage we intervene. After watching sheep give birth for many years we learn the warning signs and this is a time when our instincts become very useful.

Sometimes we just look at a ewe and know things are not right. Maybe the way she is standing or behaving is different from a normal birth. Often it is obvious. A ewe may keep lying down and pushing without any signs of a lamb, or perhaps the lamb is there but might be coming out awkwardly, not head and front feet first.

The ewe in the photograph is having a very big single lamb, which despite being presented correctly has got wedged in the birth canal. The ewe, having pushed on her own for over half an hour without success, requires some help, so the shepherd is gently easing out the lamb, or as we say he is 'lambing her'.

The shepherd doesn't need to examine this ewe internally as the lamb was partly out. Before internal examinations, it is a good idea to make sure hands are washed, rings removed and both gloves and gel applied, to cut down the risk of infection and damage to both animal and human.

This birth is straightforward, but getting some lambs out can be hard work and take a lot of time and physical effort. The less we need to examine a ewe internally the better, as it cuts down the risk of infection and damage to her. The more we can let them give birth naturally the better. But often, if we didn't help, lambs would die and eventually so would the ewe.

To some people the thought of having to put their hand in a ewe and deliver a lamb is quite horrifying, me included before I became a shepherd. I remember being really scared of that bit when I did my first lambing job. But the satisfaction of getting out a live lamb soon takes over. It is like putting your hand in warm jelly, and the bonus is you get to pull out all sorts of individuals, hopefully alive.

Scanning is really useful when we are delivering lambs. It can guide us as to how many lambs a ewe is expecting. The ewe in the photograph was scanned as carrying one lamb, so the shepherd knows she is unlikely to have any more inside her. Also the size of this lamb tells us it is doubtful that she is carrying any more; he's quite big. It is possible to feel her tummy, pushing it from the outside, to detect if she has got another lamb inside.

And he's out!

After some careful manipulation, the lamb is delivered. Both ewe and lamb are fine.

Clearing the Mouth

Once the lamb is out, I clear away any fluid from its face to ensure it can breathe. Often lambs are born with the water bag still over their face, meaning that unless we are there to clear it away they will suffocate. The lamb might have difficulty getting its breath, particularly if the head is a bit swollen, so I hold the mouth open for a while like the shepherd is doing in the photograph.

Lambs have a little sneeze soon after birth, which clears their airways. If they are really blocked up, I stick a clean piece of straw very carefully up the nostril. This tickles and encourages them to sneeze, clearing any mucus.

If lambs are still struggling to breathe they might start thrashing about. In this case I lift them up by their back legs and drain out any excess fluid. Occasionally I give them a gentle swing to clear stubborn blockages, which generally does the trick.

Some lambs can be weak and need a bit of reviving. We don't really need to dry them off – the ewe will do that – but occasionally I rub them with straw to get them going.

There are several other complications we come across when ewes are giving birth. Some ewes will have a small pelvis, meaning that giving birth to big lambs can be very difficult. In some circumstances the only way to get the lamb out is by caesarean.

Ringwomb is a condition which occurs when the cervix fails to open. Sometimes the cervix is slow to relax and it may be the ewe is not ready to lamb and just needs more time. If the water bag has been out for some time and the cervix still refuses to relax, I give her an injection of calcium. After half an hour or so, I work the muscles with my hand and they usually relax. A real case of ringwomb, however, can only be overcome by a caesarean. These are often difficult cases to gauge and it is sometimes hard to know how long is long enough.

Lambs can occasionally be born physically or mentally deformed. It is our job as shepherds to assess the severity of these conditions and administer euthanasia accordingly.

Recently we've been faced with a new disease spread by midges called Schmallenberg, which causes physical and brain defects in developing lambs. Affected lambs are very difficult to get out of the ewe, as their limbs fuse together during development in the womb and they lose all flexibility. The majority of these lambs die soon after birth. It is a really nasty condition and distressing for both sheep and shepherds.

Some sheep, especially the older experienced mothers, get very maternal before giving birth. They are so keen to have a lamb they are quite happy to steal anybody's and usually the fresher the better. I call these 'aunties'. The paler-faced ewe with the red ear tag in the photograph is an aunty.

They often get broody a few hours prior to giving birth, but sometimes it can be as early as a day or two before. These sheep are extremely persistent, disruptive and annoying. They go from one ewe to another in an attempt to find themselves a lamb. Aunties bond so closely and quickly to a lamb it is often hard to tell at a glance which mother is the real one.

The poor mother in the photograph is about to have her second lamb. She is being pestered by the thief and cannot relax enough to give birth. If she were to lie down she knows the aunty would rapidly entice the lamb away. At this stage of its life the lamb is happy to go to any ewe. When a ewe licks a lamb continuously, it will begin to smell like her and if a lamb is stolen for too long, there is a risk the real mother will eventually reject it.

In this situation I usually remove the birthing mother and her lamb from the pen to a place where she can continue lambing quietly on her own.

Handling Ewes and Lambs

When a ewe needs to be moved, we coax her using the lambs as Zoe is doing in the photograph. It is important not to go too fast or far ahead with her lambs. If she loses contact either through sight or smell, she'll panic and run back to where she gave birth.

As you can see, Zoe is holding the lambs by both front legs. This is the correct way to carry lambs and does not harm them. They can be manoeuvred easily and the ewe is still able to see and smell them when they are low down like this.

Sheep Communication

Sheep communicate using a wide range of vocalisations. Learning the different meanings to each type of bleat is very useful for us and can alert us to potential problems we often cannot see. Lambing sheds are usually full of sheep chatter.

Ewes are probably noisiest at feeding time or when they are separated from their lambs. The ewes in the trailer are being taken out to the fields. Their lambs are in the front compartment; you can see the ewes are distressed and calling to them anxiously.

The cry of pain given by some ewes during labour is also very distinct. (It always amuses me when people say to me, 'Does lambing make you feel broody?' They've obviously never been in a lambing shed for long enough!) Having given birth, ewes will rumble away softly to their babies. It's a very soothing sound.

Lambs bleat to their mothers very soon after being born. They continue to communicate their hunger, pain and anxiety through different vocalisations. When moving freshly born lambs as Zoe was doing on the previous page, we mimic the bleat of a lamb, enticing the ewe to follow. We sound pretty stupid, but it does usually work.

Treatment for Newly Born Lambs

Having made it through five months in the womb without contracting a disease and survived the lambing process unscathed, the lamb is now bombarded with life on the farm, and a whole host of other things to get used to.

It will probably have its navel sprayed or dipped with an iodine solution like the lamb in the photograph. The fresh navel of a lamb is an ideal place for bacteria to enter. Diseases such as 'joint ill', a debilitating arthritis-like condition, can develop. The iodine helps reduce this risk.

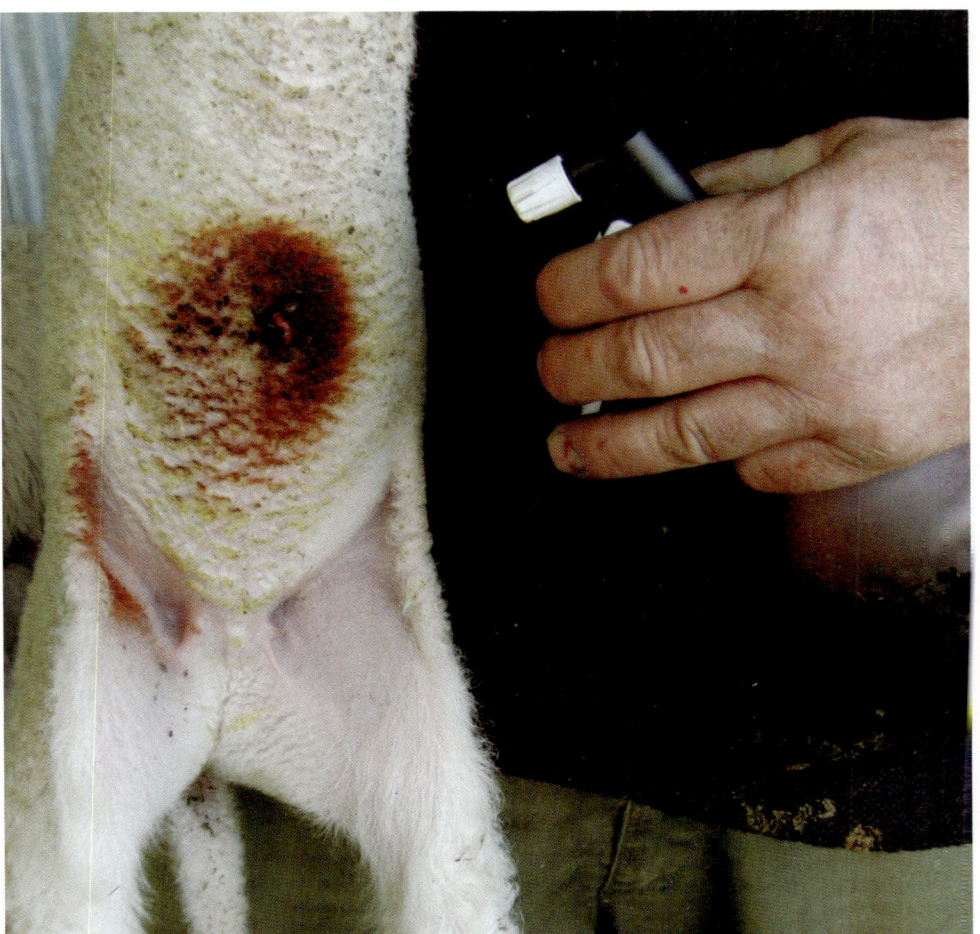

Once a ewe has given birth, she is put into an individual pen with her lambs for a day or so, like the ones in the photograph. As you can see the pens are made up of metal or wooden hurdles. Each pen has hay and a water bucket, really important for lactating ewes; they are usually desperate for a drink just after giving birth. These ewes are also fed concentrated food twice a day.

This individual pen stage is really important in my opinion and should not be rushed. Not only does the ewe need time to bond with her lamb and recover from the birth, but more importantly the lamb needs time to feed on colostrum, sleep, be warm and sheltered, and preserve valuable energy by not moving around too far. Staying in these pens for a couple of days before going to the fields, I find, reduces the number of post-lambing problems, like lambs getting lost or not feeding properly. These problems can be very labour-intensive to rectify once the sheep are outside.

If lambs are weak or struggling to feed independently or are perhaps fostered lambs, they might stay in an individual pen for several days. We female lambing staff often get the job of tending to ewes and lambs in the individual pens. It requires an enormous amount of patience and lots of fiddling about.

Colostrum

Newborn lambs need to drink as much colostrum as they can in their first few hours of life. Colostrum is special rich milk or 'first milk' packed full of proteins, energy and essential antibodies. Unlike human babies, lambs do not receive any immunity from their mothers whilst in the womb, and are born without any protection against diseases. Colostrum is vital for their survival.

Lambs can only absorb the antibodies in the colostrum efficiently on the first day after birth. Nature's balance, incredible as ever, ensures the ewe stops producing colostrum usually a day or so after giving birth, when the lamb no longer requires it. She goes on to produce normal milk for another few months.

Sometimes a ewe will not produce enough colostrum to feed her lambs and they might need 'topping up' by us for a while, to ensure they receive enough. Artificial colostrum replacements in the form of milk powder can be used. Many farms use cows' or goats' colostrum (not milk) as a replacement. It can be frozen and stored.

Are the Lambs Feeding?

Our job as shepherds is to ensure each lamb has drunk enough colostrum and is feeding independently. We spend hours going from pen to pen checking each lamb is full. Well-fed lambs are usually quietly content. They feel heavy when picked up and look plump. They will wake from a sleep, stretch out and run to the teat.

Hungry lambs, however, will at first bleat persistently. They'll stand hunched or 'pinched up' as we say and after a while they feel light when picked up. Their mouth and ears might become cold and if they get too weak they will lie in a corner and eventually die. It is vital to keep their energy levels up.

Sucking a teat, either real or artificial, doesn't come easily to all lambs. As you can see Elizabeth has to teach the lamb in the photograph how to suck the teat on the bucket, the same way we teach lambs to suck a ewe's teat if they are struggling. The younger the lambs are put on the teat, the quicker they learn to suck.

Lambs can struggle to feed for several reasons. Some ewes have low udders or large teats too big for the lamb's mouth. Some breeds have a heavy fleece which hides the udder and others, especially first-time inexperienced mothers, can be anxious and often won't stand still to let lambs drink. Some of these young mothers will actually be afraid of their newborn lambs and try to escape.

I have found moving ewes and lambs too quickly after a birth disorientates the lambs and interrupts their natural feed cycle. The more time lambs are given to stand and find the teat before any human intervention, the less problematic the feeding process seems to be.

Tube Feeding

If lambs are too weak to suck from a ewe or refuse to suck a bottle, they are tube fed. A specially designed plastic tube can be put on the end of a syringe like the one in the photograph. We slide the tube down the lamb's throat until it reaches its tummy. We pour warm milk in the syringe and let it go down gradually.

Washing the tube between lambs reduces the spread of infection, and primarily warming the tube in the milk avoids damaging the soft tissues in the lamb's throat. The trick is to get the tube in the stomach and not the lungs. Scary to start with, but you get used to it.

People often ask me how much lambs should be fed at each feed. I feed according to the size of the lamb. When I tube feed (or bottle feed, for that matter), I will look at how the stomach is filling up rather than using a strictly measured quantity of milk. A big lamb like the one in the photograph will need at least three syringes full at each feed, whereas a small lamb will need much less.

Little and often is best, ideally every three or four hours. It is vital they have enough energy to keep warm. When lambs feed from their mothers naturally, they take in far greater quantities than we will feed them over the same period of time.

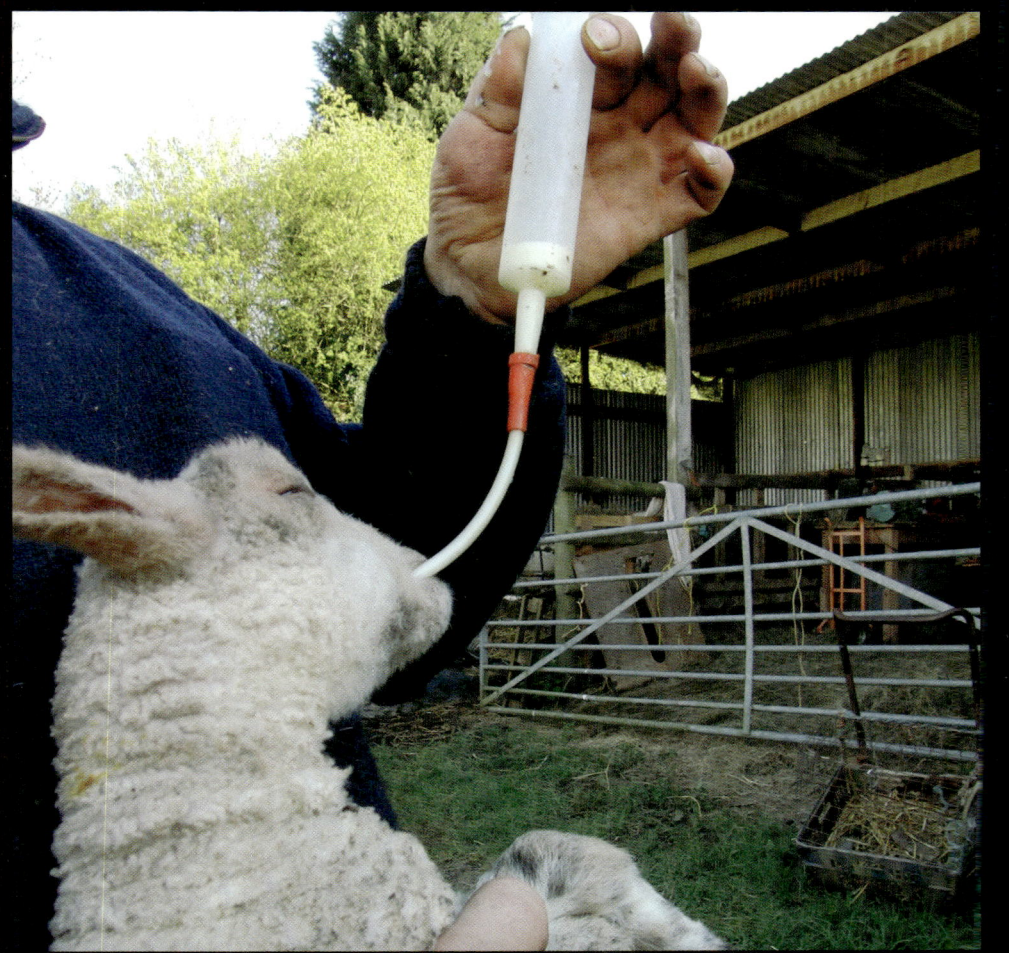

Warming Up Lambs

When lambs get really cold and run out of energy, their risk of dying from hypothermia is very high. Colostrum gives them valuable energy to keep warm, but if they are not drinking or have been exposed to bad weather during birth, for example, they will be at risk of hypothermia. A lamb's temperature can be taken by inserting a probe into its bottom, but often just looking at it or feeling if the mouth is cold will be enough. The temperature should be 39–40°C.

Hypothermic lambs that are starving react best firstly to a syringe of glucose injected straight into the stomach, followed by warmth and then colostrum. Some people use hot boxes especially designed for heating up lambs.

Not all farms have a hot box, so I developed a simple, portable way which works for me. I fill a container like the one in the photograph with straw, putting a shallow layer of very hot water in the bottom. I put in the dried-off lamb, leaving out just the head. The hot water will keep the straw warm for ages and the container can be put back with mum, to ensure the ewe and lamb bonding continues.

Once the lamb has warmed up it can be fed colostrum and should soon begin to show signs of recovery. This lamb made a full recovery.

Although many say heating lambs too quickly is incorrect, when a lamb has got very cold in the first few hours of its life, I might give it a bath (particularly if there is no hot box available). It has worked for me on numerous occasions. Lambs at just a few hours old still have an energy reserve from when they were born and will cope with the warm bath water.

Warming Up Lambs Using a Lamp

The ewe and lamb in this photograph have a heat lamp in their pen (visible in the top left corner). The cage-like wire around the base of the lamp prevents things getting burned, particularly the ewes. This lamb is now clearly doing well. He was born quite small and has undoubtedly benefited from the warmth given off by the infra-red heat of the bulb.

Some nights get incredibly cold. I have worked when snow comes in through the barn ventilation and the milk freezes in the jug before you get to feed it to the lambs. These nights are really hard work.

Some breeds cope better than others in these temperatures; some have thin coats and battle with the cold. Small, weak lambs or triplets often struggle to keep warm without the heat of a lamp. Lamb losses in these conditions can increase without adequate equipment such as lamps. There are also specially designed little lamb coats which can help to warm up cold lambs.

Fostering Lambs

ostering is when we give lambs from one ewe to another. The ewe in the photograph has been given a foster lamb and sprayed with an 'F' for 'foster', so the shepherd can identify and observe the family carefully (we'll look at the reasons for such identification later). She has been kept in this small area for a couple of days to ensure she has fully bonded with both her own and the foster lamb. This fostering was successful.

We foster to ensure every ewe capable of rearing a lamb does so and is a productive part of the farm. Fostering also means a natural life and higher chance of survival for a lamb, as opposed to being artificially reared.

There are many circumstances in which we foster. Ewes often have triplets. Supporting triplets can drain a ewe, unless she is fed supplementary feeds, and will often compromise the growth of all three lambs. Generally the smallest or weakest lamb will be taken away and fostered.

Sometimes ewes and lambs die, creating fostering opportunities. Ewes that have one lamb of their own but enough milk to rear two can be given an additional lamb to rear. And conversely, lambs born to a ewe without enough milk may be taken away and fostered on to another mother.

Certain breeds accept foster lambs easier than others. Some ewes are extremely switched on to this game. I find Dorsets are great to foster onto, whereas mules can be awkward. Most ewes want to be mothers and will accept a single lamb without too much persuasion. They are often fussier if given the choice of two lambs to care for, particularly if one is their own and one a foster. A dog can come in useful when fostering. If a dog is in front of the pen the ewe's maternal instinct will increase and she usually becomes very protective, even of a lamb that is not her own.

Fostering is tricky business. It tests our shepherding skills and can take time and effort, but when it works it is very rewarding. We will look at the different methods of fostering next.

Wet Fostering

Wet fostering is one of the most successful and easiest ways to foster and is done at the time of birth. The ewe in the photograph was scanned as having a single lamb but has enough milk to rear two, so Matt is fostering another one onto her. The lamb on the right is the mother's natural lamb and the one on the left is a foster.

The foster lamb here is smaller than the ewe's own lamb. It is ideal, although not always possible, to match the two lambs in size so that one doesn't have to compete for milk. However, we do often find these smaller lambs can be real fighters and need much less milk than bigger lambs. Providing they are strong enough, they will generally keep up. This situation doesn't arise solely with fostering; ewes will occasionally give birth naturally to lambs of uneven size.

When I wet foster, I get a warm bucket of water and put in as much fluid from the ewe as I can. If I'm fostering onto a single ewe like this, I put both her own lamb and the foster lamb in the water so they smell the same (Matt is putting bits of broken water bag over the lambs to disguise any 'foreign' smells). I then present the two lambs to the mother.

Occasionally, depending on the circumstances, I might take the ewe's own lamb away and let her bond with the foster first. Providing her own lamb isn't away from her too long (not more than 15 or 20 minutes), she will accept it back. This foster was successful.

Skinning

Skinning is when we take the skin from a dead lamb and put it onto a live, healthy one. The little chap in the photograph is wearing a skin. The skin blends in very well and you can just see the baler twine under his neck securing his new coat.

I match a potential foster lamb to a ewe carefully to maximise the chance of success. The ewe may have been with her own lamb for a while before it died and got to know its characteristics. Suddenly putting her with a different lamb, and often a livelier one, doesn't always work.

The head and legs of introduced lambs with skins on are exposed and will not smell like the ewe. There are sprays you can use to mask the scent of newly introduced lambs, but much depends on the temperament of the ewe as to the outcome.

After a few days of drinking milk, the lamb will smell more like the ewe and the skin can be taken off. I do find these fosters are worth persevering with even if they seem unsuccessful at first.

This foster worked very well; the ewe adores him. She was stamping her feet protectively whilst I took the photograph. Generally most sheep stamp their feet to protect their lambs, although there are the headbutters and I have come across the odd spitting and hissing sheep.

Lamb Adopter Pens

If wet fostering and skinning are not suitable options, then an adopter pen like the one in the photograph can be used. People are often upset when they see ewes confined like this, but sometimes it is necessary for both ewe and lamb. I have seen ewes severely injure and kill lambs they dislike, hitting them with their heads.

The ewe is free enough to get up and lie down, eat, drink and rest comfortably, but restrained enough to ensure she cannot see or harm the lambs. This method keeps lambs safe. The idea is that in a couple of days the lamb or lambs smell like her and a bonding will have taken place.

Not all ewes bond naturally with their own lambs. Occasionally a ewe will dislike one or more of her own lambs. These ewes are difficult. There is often little we can do to change their minds, but it is worth trying the head lock for a few days.

The earlier you catch the problem and head lock a ewe, the better the chance of success. After three or four days, if she still dislikes the lamb, it is unlikely she will ever change her mind and I find it is better to abandon the idea. I feel it's important to work 'with' a ewe to avoid breaking her spirit. Assessing the circumstances and knowing when to call it a day is better for the ewe, lamb and shepherd.

Once we are satisfied the whole family is doing well in their little pen, there are several procedures, including tailing, castrating, numbering and possibly ear tagging, which need to be carried out on the lambs before moving them on to the next stage.

Most lambs are born with long tails. Tails protect sheep's bottoms and udders from bad weather. But sometimes, especially for adult sheep, long tails get in the way. Lowland-kept sheep in particular, eating a rich grass diet, will often have runny bottoms. Their tails get mucky, which attracts the flies and can lead to a nasty condition called flystrike. So it is desirable and common practice to reduce the length of the tail. Short tails also make shearing easier and are necessary for showing certain breeds.

It is against UK welfare regulations to dock tails really short, which would leave the animal exposed to the elements. I have also noticed very short-tailed sheep seem to be more prone to cervical prolapses.

There are several methods used to reduce the tail length, but the most common way is to use a rubber ring. A ring is put on the tail like the shepherd is doing in the photograph. We call the procedure 'ringing'. In a couple of weeks the unwanted end will wither and fall off.

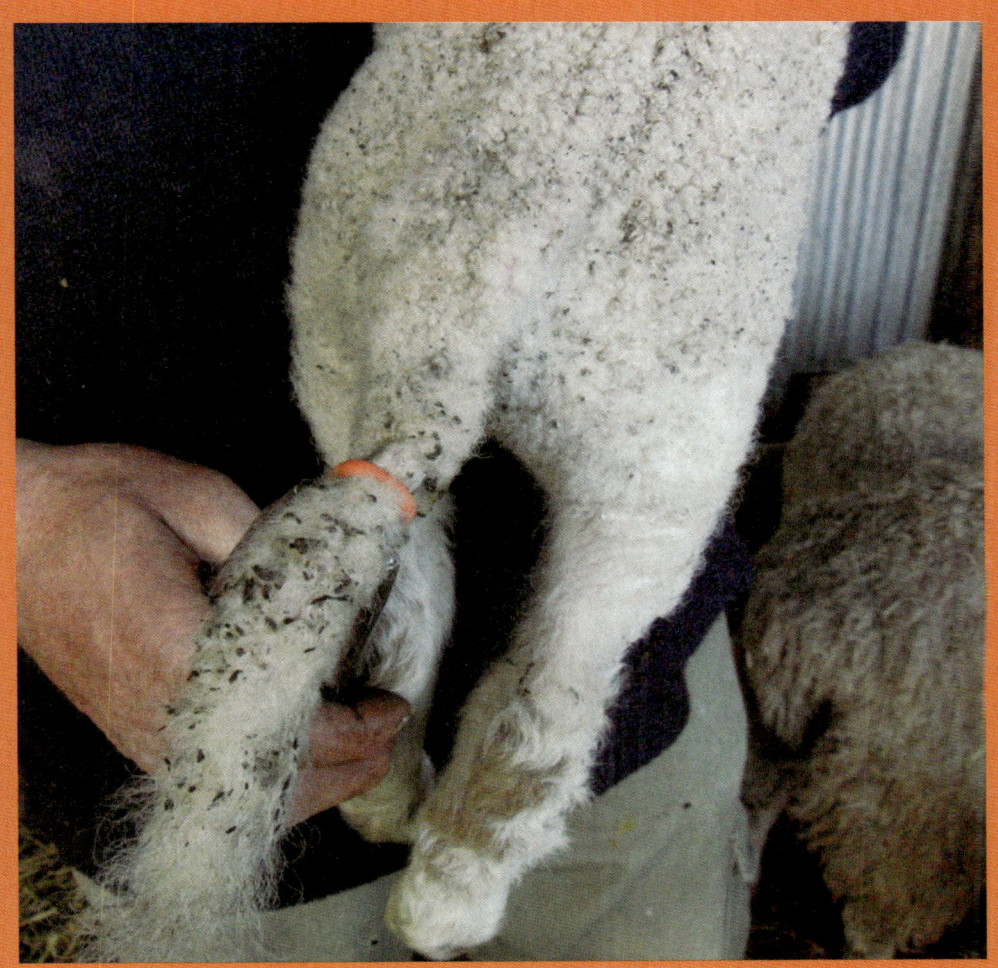

Castrating Male Lambs

Ram lambs in commercial UK flocks will often be castrated. Lambs are castrated for several reasons.

Male lambs are usually destined for the butcher. It is often believed an uncastrated lamb will not produce as good-quality a carcass as a castrated lamb, particularly if they live longer than five or six months. (Uncastrated lambs start running around after each other, acting and growing like rams, and it is harder to fatten them.) It is also said the meat of older uncastrated lamb is tainted in taste. Then of course there is the fertility problem. At five or six months of age, uncastrated lambs will become fertile and need to be separated from ewes and female lambs.

There are several methods which can be used to castrate male lambs. The most common way, as with the tails, is to put a rubber ring around the scrotum with callipers and rubber rings like the ones in the photograph. The scrotum will shrivel and fall off in a few weeks.

The procedure of ringing is controversial as it causes pain, particularly when used for castration. It is especially painful for older lambs. In my opinion, if a lamb needs to be 'rung', it is optimal to carry out the procedure between one and two days of age to keep the level of pain to a minimum (providing the lamb is strong and well fed and has had plenty of colostrum). By UK law, if a lamb is to be ringed without anaesthetic, it must be done within a week after birth.

Numbering Ewes and Lambs

We need a way to identify each animal when it leaves the shed, so once lambs have been rubber ringed they will usually be sprayed with a number.

Spraying the same number on individual family members, as the shepherd is doing in the photograph, allows us to clearly identify which ewes and lambs belong together. I suppose it's the equivalent of a surname for us. This means when lambs wander off or get separated from their mothers, we can reunite them quickly. Others might need treatment and observation if they are sick, and having a clear visible number like this allows us to identify any animal easily.

Like the ram raddle marks, marker spray is approved by the wool board and will last several weeks on the fleeces, either naturally fading or washing out when the ewes are shorn and fleeces prepared for wool products.

Ear tagging is another form of identification. Some farms choose to ear tag lambs shortly after birth, as you will see in some of the photographs. This is not required by law, but will be necessary when animals move away from the farm or reach a certain age. The tags have an individual number as well as a flock number, so that every animal in the country has its own specific numerical identification throughout its life.

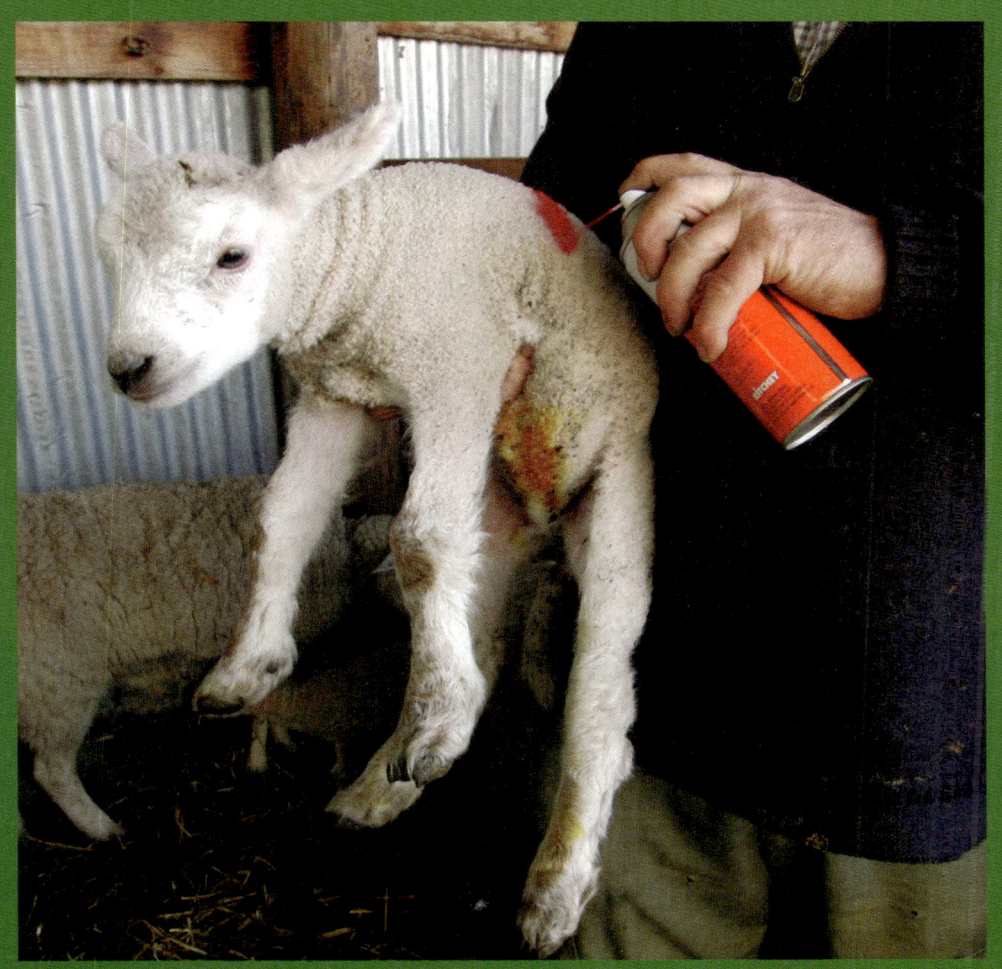

Ailments Suffered by Newborn Lambs

L ike their mothers, lambs are susceptible to many diseases and conditions, and as I mentioned previously they are particularly vulnerable immediately after birth when their immunity is fragile. Lambing sheds are a fabulous playground for both good and bad micro-organisms. Keeping a shed clean will help reduce the lambs' risk of infection. Here I will explain a few of the common conditions affecting young lambs that we see during lambing time.

Watery mouth is a particularly nasty disease and lambs risk developing it during their first few days of life. The E. coli bacteria responsible for this condition exist everywhere in the environment, but become more dangerous to lambs when their intake of colostrum is compromised. Watery mouth can appear subtly at first and it's easy to mistake the symptom of a bloated tummy for a full one. Infected lambs drop their head and ears and they drool saliva from their mouth, hence the name 'watery mouth'. They are unable to swallow due to a stomach full of gas. The disease develops rapidly and unless treated, lambs die quickly.

Working continuously in the little pens certainly helps one catch this disease early. There is a section at the back of the book explaining the way I treat this disease.

Joint ill is another nasty disease affecting lambs in their first few weeks of life. It is an incapacitating arthritic condition affecting the joints, particularly in the legs. It is triggered by various bacterial strains. The bacteria enters the lamb usually through a wound or, as I mentioned previously, the fresh navel.

Dirty lambing pens and a lack of colostrum also increase the risk of lambs developing this disease. This condition doesn't always show up in the little pens and generally lambs have gone outside before it develops and we see them limping in the fields.

When I find a case of joint ill, I administer an antibiotic and an anti-inflammatory. Treatment needs to be intensive and ongoing to get this condition under control. Treating a lamb on day one is quite simple as they are obviously in pain and don't move very quickly. The more they improve, the faster they become, and it can be really hard to catch them towards the end of their treatment.

E. coli, salmonella and rotavirus are all conditions which can affect young lambs and give them diarrhoea. Again dirty conditions and lack of colostrum contribute towards these outbreaks. Lambs are also susceptible to vitamin deficiencies and sometimes I will administer a multivitamin injection to cover all eventualities.

Inverted Eyelids

A condition we may well see is inverted eyelids or entropion eye. The lower eyelid turns inwards and rubs continuously on the cornea, causing the eye to run and blindness if left untreated. I have seen this condition more frequently in certain breeds, particularly white-faced sheep, and it is said to pass down from the rams.

In a busy lambing shed it is sometimes easy to miss this condition, particularly if the eye is only mildly affected. But in severe cases, particularly if it occurs in both eyes, inverted eyelids will stop the lamb from drinking, following its mother and behaving normally, mainly because it can't see!

I can only imagine how painful this condition must be, but it is usually quite straightforward to fix. If the case is only mild, turning out the eyelid to its correct position is sometimes enough, but if not I inject the eyelid with an antibiotic diluted with water. Using a small needle I inject enough solution to puff out the eyelid and prevent it from turning back in. The eyelashes stay out of the eye and the antibiotic kills any infection. This procedure usually only needs to be done once and rarely fails to correct the problem. If, however, the case is severe, then stitching the eyelid back may be the only solution – one for the vet!

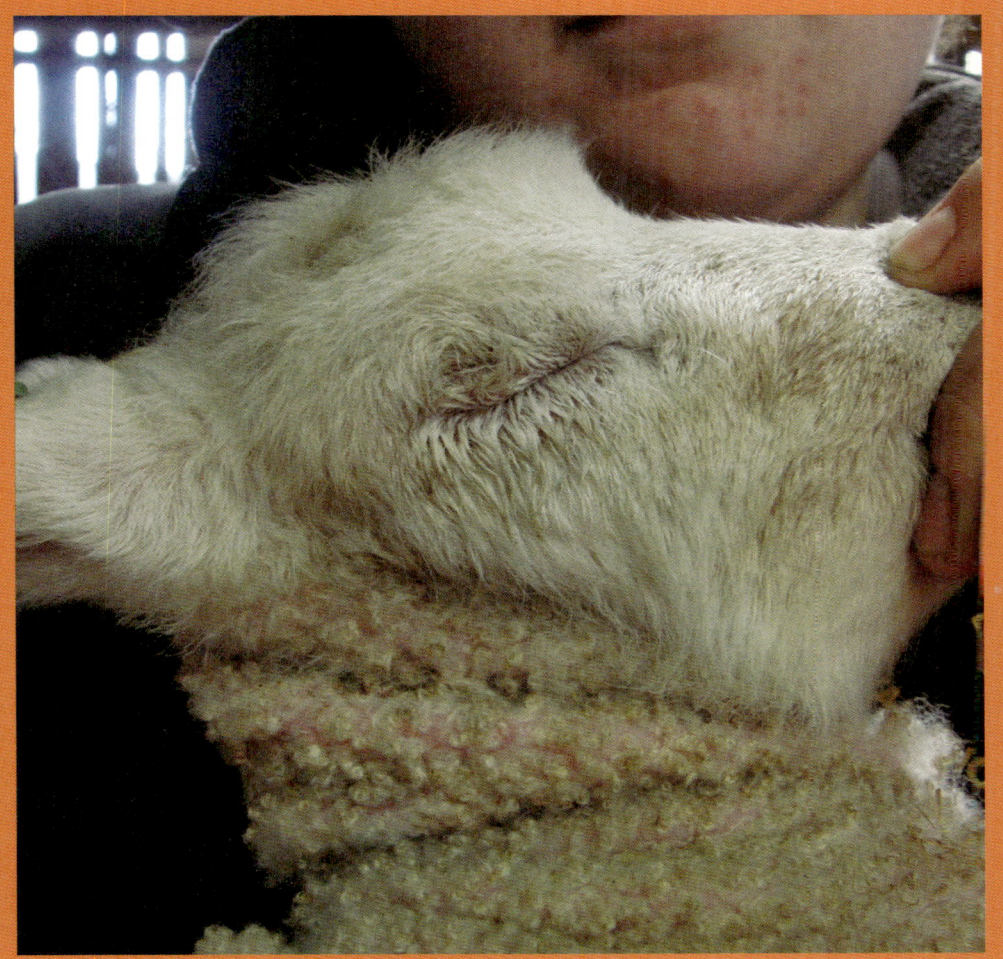

Contracted Tendons

A contracted tendon is a disorder we often come across. Basically, the tendon is too short for the leg and pulls the toe backwards. It means the lamb walks on the knuckles, which is not natural and must be uncomfortable. Many animals including foals, goat kids and calves can also suffer from this problem.

Genetic factors are thought to contribute to this condition. Sometimes a contracted tendon can right itself after a few days, but not always. If it is caught early, I have good success using a simple piece of cardboard and a technique shown to me by a shepherding friend.

As you can see, the lamb in the photograph has a tube of cardboard taped around her front leg. I wrap a layer of cardboard (corrugated, I find, works the best) around the leg and tape it together, making sure the tape is not too tight. This allows the leg to grow unrestricted. As the knee joint is kept free, she can get up and down easily and still run around. It usually takes about a week for the condition to come right.

There are many challenges we face at lambing time and it is our job to make sure a sheep's life is as good as it can be, however long or short it may be.

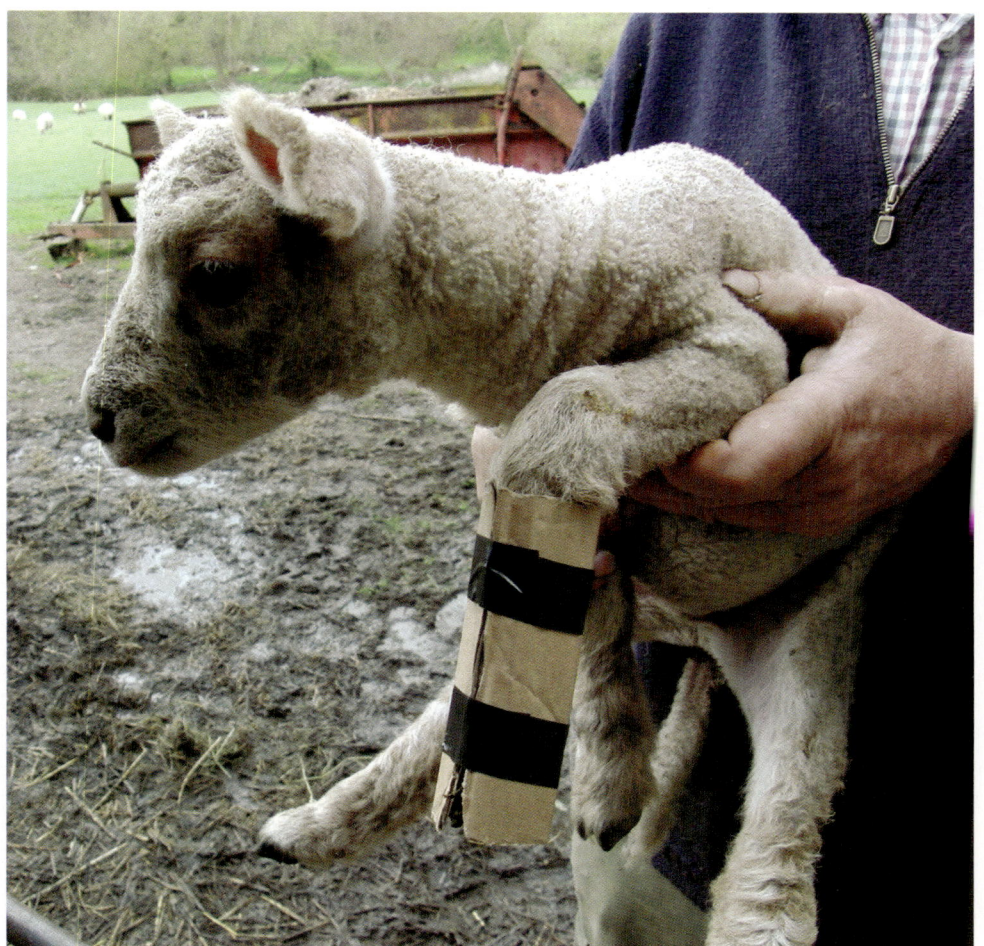

Artificially Rearing Lambs

Depending on where you live in the country, artificially reared lambs are referred to by many different names. In some places they are known as cade lambs; others call them Mollies, pets, socks or, if you're like me, orphans. (Although this isn't strictly accurate, as apart from the odd lamb, most haven't really been orphaned.)

An excess of 'spare' lambs usually comes from taking lambs away from triplet ewes or ewes without enough milk. At one farm I work on, the ewes will often have more than two lambs each (the largest number of lambs I have seen born live to one ewe is five). This farm has large numbers of orphan lambs to rear each year and they use an automatic milk-dispensing machine, like the one the lambs are feeding from in the photograph.

The most delightful thing I saw this year was putting footballs in with groups of orphan lambs. The lambs push the footballs round with their heads. It's very funny and keeps them, and us, entertained for hours.

Each shepherd takes a different view on orphan lambs. Some avoid them at all costs, fostering when they can or selling any spare lambs on. Others see orphans as an inevitable fact of lambing and set up a system to house and feed them. They can be very time-consuming in the beginning, are quite expensive to rear, as milk powder isn't cheap, and have less immunity than naturally reared lambs. But if they are kept clean and have the correct diet, orphans can grow into good healthy sheep.

What Do Lambs Eat?

These two-week-old artificially reared lambs are experts at the milk-drinking game and have progressed to eating solids in addition to their milk.

After a week or so of drinking milk, orphan lambs can be introduced to solid foods such as hay and special pellets called creep, which the lambs in the photograph have in their trough. The solid food helps with the development of their stomachs. They will also be introduced to drinking water at this stage. (Most lambs that are reared naturally on ewes are not introduced to creep until much later, if at all. These lambs tend to live just on ewe's milk and grass for several months.)

Most orphan lambs are weaned off milk at about six to eight weeks of age. Some will wean their lambs in accordance with their weight rather than their age, but this varies from farm to farm. Only once lambs are eating creep and hay can they be weaned off milk. They will start nibbling grass when they are about ten days old.

Do Lambs Attach Themselves to Humans?

Yes, lambs attach themselves very quickly to humans, particularly lambs reared on a bottle. This connection stays all the time we hand feed them, regardless of the procedures we carry out on them. Lambs will, however, always choose to be with a ewe in preference to a human, which is of course how it should be.

Lambs don't stay cute for very long and after a couple of months, the adorable lamb in the photograph will have changed completely. I totally appreciate the non-meat eater's view, but I feel sad to think without a meat market there would be very little need for these lambs to be born at all.

Do Humans Attach Themselves to Lambs?

We usually have a favourite amongst the lambs (especially us girls), but we only care for them and understand they are not pets. They will grow up very quickly and no longer need us.

There was a special lamb in my life once called Alice. She had a white face and looked very much like the lamb in the photograph. She was born on a neighbouring farm. Her mother had died so I took her home. She contracted meningitis and had to stay away from the rest of the flock. During her recovery, she'd ride round in the pickup with the dogs; they were very protective of her. This used to cause quite a stir with passers-by when we'd stop off at the shop.

When Alice made a full recovery, she went to live amongst a friend's flock. Whenever I visited she would leave the others to come and see me. I never fed her treats; she came because she wanted to.

She went on to have many lambs of her own over the years and one spring whilst I was watching her in the fields, she left the flock, followed tentatively by her two lambs. She walked over to where I was sitting and lay down beside me. This act reassured her lambs that this human was 'OK'. I felt it a great privilege that she had trusted me enough to show me her new family in this way. We had a very gentle relationship formed over many years. I will never forget her.

Time to Join the Flock

This photograph is one of my favourites. It truly was a serene sight, lambs so content, peacefully soaking up the sunshine.

Any ewes with lambs which are not ready to go straight from a little pen to the fields can be moved to an open communal area, like the one in the photograph. Here they can sleep, drink from their mothers, play, socialise and explore without too much fear.

Often very small lambs or lambs that have been ill will be held back in these communal or group pens for several days until they are fit enough to go outside into the fields. The lambs go to the field stronger and more 'streetwise', having experienced a bit of life.

The ewe and lamb bond is very strong from birth and generally remains so as the lamb grows older. Ewes rarely tolerate other lambs drinking from them and will only stand to let their own lambs drink. Lambs will not usually stray to another mother unless the milk dries out. If lambs get hungry they go hunting for a full udder, often sneaking in behind any ewe to get a free meal.

Group Pens

These ewes and lambs are waiting to go outside. They have been in this barn for a couple of days now. As you can see they have clean straw bedding and an adequate supply of good-quality hay. They are also fed concentrated food twice a day. Some farms continue feeding concentrates to the ewes after they are turned outside to help with their milk production.

When lambs reach a few days of age they love to climb and play. They like climbing up into the round hay feeders and often race round in gangs.

One farm I worked on for several years would lamb in February and the ewes and lambs were kept in big group pens for a week or so to ensure they were strong enough to cope with any bad weather before going outside. Around the same time each evening, the lambs would gang together in a big mob and race up and down the barn. It was hilarious, especially watching the younger ones trying to keep up at the back. They continue to play like this for several weeks. It really is great fun to watch. We occasionally get 'playtime' injuries, usually limping legs!

Some shepherds like to turn ewes and lambs out into the fields as quickly as possible and don't use group pens like we saw on the previous page. Some farms just don't have enough undercover areas to house ewes and lambs and have to turn them out into the fields as soon as they can.

Much depends not only on barn space, but also on the weather. Wet, cold conditions are the worst thing for newborn lambs. Warm sunny weather is what they need. We make the most of sunny days and turn out as many families as we can.

Unfortunately we can't change the weather, but providing lambs go out strong, they will cope with most conditions. Some shepherds put out straw bales for lambs to shelter against. (The ewes are fine and don't feel the cold; they have nice wool coats!)

The quad bike and trailer in the photograph are great for transporting small numbers of ewes and lambs. There is a compartment in the front of the trailer where lambs go, preventing them from getting squashed in transit. Lambs' legs can be fragile and will get broken if they are squashed or trodden on. They are especially vulnerable during turning out. Broken legs in both lambs and ewes can be fixed with plastering, providing the break is clean.

Turning Out with a Trailer

A large trailer is useful for transporting lots of ewes and lambs. When they are ready to go to the field, the lambs are loaded up in the front compartment or top deck of the trailer so they don't get trampled, whilst the mothers travel either in the back or underneath.

There is often much confusion when they reach the fields and are let out. It is ideal if ewes and lambs are taken out to the fields in the morning. This way they have all day to regroup within their individual families or 'mother up' as we say, and the lambs have time to adjust to their new environment before it gets dark. If we can, we return to the field, making sure everybody has mothered up and is settled before nightfall. For hill farms and farms that have large numbers of ewes, things will probably be very different.

The lambs in the photograph are tentatively making their way down to the bottom of the ramp, where some of the mothers are waiting. Ewes are used to riding in the trailer, but for the lambs it is a whole new experience.

The Beginning of a New Chapter

It is a great relief when we get sheep returned to the fields, but this is when the hard part for shepherds begins: trying to keep everything alive and healthy. Magnesium deficiencies, mastitis, joint ill and a whole host of other obstacles are suddenly there to overcome.

During lambing, each person tends to have their own role. I am usually shed-based and don't often get to go round the fields until the end of lambing. Sometimes I might only get to see the lambs I have cared for after two or three weeks. Seeing strong healthy lambs in the fields makes the job of being a shepherd at lambing time very rewarding. This is the part which makes all the hard work worthwhile.

Another Lambing Over

So there we are. The shed is almost empty and most of the ewes are out in the fields enjoying the spring sunshine with their new families. One more lambing over for this farm, but somewhere there will be another farm and another flock of pregnant sheep waiting.

I hope you have enjoyed reading this book and maybe discovered some things about sheep and shepherds. Lambing is a very significant event in the shepherding calendar, but it is still only a small proportion of a sheep's year. The rest is another book entirely.

My Treatment for Watery Mouth

For watery mouth cases, I administer a broad spectrum antibiotic and an anti-inflammatory (if available). I tube-feed the afflicted lamb an electrolyte or diluted high-energy liquid if there is room in the bloated stomach. Tubing not only provides a source of energy but fills the stomach, preventing lambs from drinking the ewe's milk. Milk accelerates this condition and the lambs must avoid it until they have recovered.

I am often asked at what stage milk can be drunk safely again by the lamb. Each lamb is different and I have found the severity of the case will usually determine the recovery time. A lamb caught early with the disease will recover quickly and is usually ready for milk after about a day. Much sicker lambs will take longer to recover.

If I think a lamb has recovered, i.e. if the bloating has gone and the drooling saliva dried up, I start by feeding it a tube of milk (that is if it hasn't already started drinking from the ewe, which is a possibility as it will be feeling much better). If after feeding the saliva returns, I know the disease is still active and I go back to the antibiotic and electrolyte/high-energy liquid routine. This way works well for me.

Watery mouth is obviously distressing and detrimental for lambs. Reactions to treatment can be slow; administering treatment is time-consuming and greatly increases our already heavy workload. Prevention really is better than cure. Keeping a clean barn and a good level of colostrum production reduces the risk of watery mouth developing. Ways of helping in the colostrum department include sufficiently feeding in-lamb ewes and taking out of the flock any old ewes, or ewes with bad udders, before tupping.

Glossary

Crossbreeding	Mating a ram with a ewe of a different breed
Dagging (also crutching)	Shearing an area around a sheep's tail and bottom
Ewe	Female sheep
Ewe lamb	Female sheep under a year of age
Flystrike (also myiasis)	A condition caused by flies laying eggs on a sheep; the eggs hatch into maggots and the maggots eat what they can find – the sheep
Hogget	A sheep between the stages of weaning and first shearing
In lamb	(of a ewe) Pregnant
Lactating	Producing milk
Lamb	A sheep under one year of age
Lambing	(of a ewe) Giving birth
Ram	Male uncastrated adult sheep
Ram lamb	Male uncastrated sheep under a year of age
Shearing	Clipping off all a sheep's wool
Shearling	A sheep which is between its first and second shearing
Gimmer	A female sheep which is between its first and second shearing
Tup	Ram
Tupped	(of a ewe) Mated with
Tupping	Mating
Turning out	Taking ewes and lambs to the field
Weaning	(of a lamb) Coming off milk
Worming	Treating animals for worms and internal parasites

References

Abbey Veterinary Group. http://www.abbey-vetgroup.co.uk (accessed 2012)

Department for Environment, Food and Rural Affairs. 'Detailed results and datasets':
http://www.defra.gov.uk/statistics/foodfarm/landuselivestock/junesurvey/junesurveyresults/ (accessed July 2012)

Henderson, David C (1995). *The Veterinary Book for Sheep Farmers*. Farming Press UK

Hiendleder, S; Kaupe, B; Wassmuth, R; Janke, A (May 2002). 'Molecular analysis of wild and domestic sheep questions current nomenclature and provides evidence for domestication from two different subspecies'. Proceedings. *Biological Sciences*, Royal Society of London, v.269, pp.893–904

Meadows, Jennifer RS; Cemal, Ibrahim; Karaca, Orhan; Gootwine, Elisha; Kijas, James W (March 2007). 'Five ovine mitochondrial lineages identified from sheep breeds of the Near East'. *Genetics*. v.175, no.3, pp.1371–1379

Merck Sharp & Dohme Corp. *The Merck Veterinary Manual*:
http://www.merckvetmanual.com (accessed Sept 2012)

North of England Mule Sheep Association. *The North of England Mule*:
http://www.nemsa.co.uk/north-england-mule (accessed Sept 2012)

Descriptions of Sheep Meat

Lamb	Meat from a sheep less than a year old
Hogget	Meat from a sheep between one and two years old
Mutton	Meat from a sheep over two years old
Salt marsh lamb	Meat from a lamb raised on salt marshes

Acknowledgements

This book is not only a story of what happens to sheep and shepherds at lambing time, but it is also a reminder to me of all the wonderful friends I have made over the years. It was great seeing you and your farms again. I would like to say a special thank you to Tom, for your ideas and positivity, and indeed all those at Acorn Farm. Thanks to Carolyn, Tim, Chris and Julie. Thank you to FAI at Oxford, you have an incredible selection of very cute lambs. Thanks also to the beautiful estate of Overbury in Gloucestershire, whose sheep feature throughout this book. There were many more of you who supported and encouraged me whilst I wrote this book. I am truly grateful to you all.

About the Author

Carol grew up in Kent and first started working with sheep about 20 years ago. Her passion for exploring the outdoors led to adventures all over the world. She rode her motorbike in the Australian Bush, travelled across south-east Africa and ten years ago took Sam, her border collie, to New Zealand for a year. They explored the South Island in a little van. Having rapidly departed education at 16 to go travelling, Carol went back to study in 2003 and achieved a degree in Environmental Science. She currently lives in the south of England.

Carol can be contacted at: www.thesheeplady.co.uk
enquiries@thesheeplady.co.uk